The
Physics of Bodies
In Real
And Imaginary
Spaces

WILLIAM L. STUBBS

4U
(For You)

.

Contents

Preface

With all we know about the world we live in, fundamentally, it remains a mystery to us. We struggle to understand things like how forces work, why mass exists and what light is. We have built a whole system of physics that explains the bulk of higher-level phenomena based on faith that the foundational phenomena will continue to behave the way they always have. However, we do not know why they behave as they do. *The Physics of Bodies in Real and Imaginary Spaces* is my attempt to assess basic phenomena such as space, mass, velocity, energy, and momentum based on what nature reveals to us, and propose explanations for why the world behaves as it does.

Of course, my investigations of these topics are by no means the first. To some degree, people have wondered about the space we live in, the forces that control us and the stuff we are made of for nearly as long as there have been people. My inquiries and conclusions regarding these phenomena certainly do not carry the authority of other works done before mine and currently in progress. Many smart people have studied these topics over the years and continue to study them. So, at the very least, my theses are just other perspectives of concepts that continue to elude our complete understanding.

I begin by discussing how the evolution of numbers and number systems has expanded our awareness of the physics occurring around us. I talk about the initial connection between numbers and objects. I discuss how providing closure for mathematical operations such as addition, subtraction, multiplication and division expanded number systems with numbers that initially seem abstract, but were later found to have physical significance.

Using this model, I propose that the existence of imaginary numbers implies that a physical realm built of imaginary numbers complements the real world made of real numbers that we take for granted as the entire world. I propose that the universe is made of real space that I call the apparent world, and imaginary space that I call the hidden world.

From this premise, I describe the coordinate systems that define each space and how the spaces relate to each other. I talk about apparent (real) world mass and its connection to the hidden (imaginary) world through a constant hidden world velocity it possesses. I show how using velocities that have both apparent world and hidden world components naturally produce expressions of relativistic concepts such as length contraction and time dilation.

I consider the likely properties of hidden (imaginary) world mass and show that hidden world mass particles moving within the hidden world project momentum into the apparent world. Finally, I suggest how the momentum projected into the real world by the imaginary mass moving in the hidden world could be the electromagnetic radiation we see in the apparent world.

Presented here for your perusal, review, comment, criticism and enjoyment is *The Physics of Bodies in Real and Imaginary Spaces*.

William L. Stubbs
Port St. Lucie, FL
October 2016

*The Physics of Bodies in
Real and Imaginary Spaces*

1
The Physical World
In Numbers

Physics is our attempt to describe the world around us. Through observations, measurements, experiments and analyses we collect information about our world and formulate theories and models of how it appears to behave. The success of these explanations depends on how well they mimic and predict the physical behavior they attempt to model. Usually, theories of physics enjoy good acceptance for a while, but succumb, in time, to new, more comprehensive models of the phenomena.[1] Over centuries of trial and error, we have gradually developed a fair understanding of how the universe works, and, in some instances, why it works the way it does.

In developing this understanding of the world, we found mathematics to be the language of physics. Through mathematics, we can express, rather precisely, how phenomena behave in space and time. This allows us to probe aspects of nature, such as traveling at the speed of light, which we could never experience in real life. It also allows us to interpolate into, and extrapolate out to, worlds that are unattainable to us. With mathematics, we can peer into the confines of the subatomic world and know our way around; or look out with some understanding into the vastness and complexity of our galaxy and the universe beyond it. Mathematics and its continued development provide a pathway into a deeper understanding of our world.

From mathematics, we discover that numbers and number systems establish the structure of the world it describes. Throughout history, when faced with perplexing situations, the discovery of new and different types of numbers[2] shed light on the previously unexplainable. The numbers associated with various phenomena have expanded our insight and understanding of the world we live in. Numbers tell us what various aspects of the world look like, even when those aspects are not directly accessible to us. Because of this, exploring numbers can give us a deeper understanding of how our world works.

The numbers that we take for granted today were not always in place in the world. Some numbers, for example, imaginary numbers, were not recognized in some circles until as late as the 16th century, relatively recently in man's history. [3] Pythagoras, one of the greatest mathematicians of all time, did not consider zero a number, nor did he recognize the need for negative numbers. It appears people discovered numbers on an "as needed" basis.

It seems likely that the discovery of numbers came out of the necessity to manage objects.[4] Knowing how many objects one has and keeping up with their movement was, and continues to be, an important aspect of life in the world. Early in human history, the ability to count objects could have meant the difference between having enough food for the winter or not, or knowing how large your rival forces are or not being adequately prepared to defend yourself. Having numbers that correspond to objects in order to understand their magnitude, humankind possessed a crucial tool for its survival and progress.

The mathematical operations we use to manipulate numbers, addition, subtraction, multiplication, division, and others, revealed the limits of the various types of numbers we use, which led us to find undiscovered numbers. Subtracting large numbers from smaller ones revealed negative numbers. Dividing large numbers into smaller ones revealed fractions. Considering the square root of negative numbers led to the discovery of imaginary numbers. In many instances, closing operations[5] revealed the need for numbers that, at the time, seemed abstract, but in time found applications in physical situations. Objects and operations determine the nature of numbers in the physical world and led to their discovery over time.

To get a sense of how numbers may have evolved, let us consider a table with several pizzas on it. If we want to know how many pizzas there are on the table, we simply count them, one, two, three, etc. until we have counted them all. We use the set of natural numbers, numbers that align with individual objects, to do this. The need to count whole objects revealed the natural numbers, 1, 2, 3, etc. A natural number corresponds to each object we need to count.[6] We can add to the number of pizzas we have on the table and the resulting number of pizzas will still be a natural number.

We can subtract from the number of pizzas on the table and still have a natural number of pizzas left, unless we take away all of the pizzas on the table. Then, with no object left, there is nothing left to count. Consequently, there is no natural number, no counting number, representing having no pizzas on the table. This situation requires that we add the number zero to the set of natural numbers, forming the set of whole numbers.[7] Now, if we remove all of the pizzas from the table, we have a number to describe the amount of pizzas remaining. The subtraction operation revealed a shortcoming in the set of natural numbers, and the existence of the number zero for managing objects.

As we prepare to distribute the pizzas to our customers, we realize that there are more customers requesting pizzas than there are pizzas on the table. Now, each customer represents one less pizza on the table. When there are no pizzas left, each remaining customer without a pizza represents one pizza less than zero on the table. No whole number represents having less than zero pizzas. This type of situation revealed the concept of numbers less than zero. These numbers became negative numbers and the numbers greater than zero, positive numbers.

Subtracting a number from a smaller number reveals the existence of negative numbers. Having only 10 pizzas on the table but needing 20 to satisfy the customer orders leaves −10 pizzas on the table. Therefore, closing the subtraction operation expanded the set of numbers, once again. Negative numbers extended the set of whole numbers to form the set of integers. Initially, people likely recognized negative numbers as needed to close the subtraction operation, even though, for objects, they appeared to be abstractions.[8] There is no such thing as a negative pizza. However, eventually a physical application for them emerged.

Negative numbers initially appeared to be abstractions because they have no physical meaning where objects are involved. For example, you cannot eat a negative pizza. Negative pizzas do not physically exist. If we have five more customers than pizzas, after selling all the pizza we have, we do not say, "We have negative five pizzas left;" instead, we say, "We need five more pizzas." We can talk around the need for negative numbers for objects. Therefore, initially no numbers less than zero appeared to have any physical significance. They could not be associated with actual physical objects, only abstract ones.

Eventually, however, we realized that some physical situations do require number less than zero. If, for example, we label the point on a thermometer where water freezes 0, and the boiling point of water, 100, then what label should we give a temperature colder than the freezing point? Without negative numbers, there is no value for this label.

Negative numbers allow us track positions of objects relative to references. A temperature of 17 degrees colder than the freezing point of water is –17 degrees. If you are at a depth of 200 feet below the surface of the ocean, you are at –200 feet. At 10:30 AM, you are –15 minutes from a 10:45 AM launch. Negative numbers let us express, mathematically, occurrences before or behind a designate origin point. So, the closure of an operation, in this case, subtraction, appeared to produce numbers that were abstractions, but actually revealed an aspect of the world that initially was not readily apparent to us.

A group of children come to the table and asks for pizza. We cut one of the pizzas into slices and serve the kids. How much pizza did each child get? No integer describes part of a whole pizza. The need to describe parts of a whole occurs when dividing larger integers into smaller integers, such as five children into one pizza. This type of situation revealed fractions.[9]

Fractions, such as ½ and ¾, are the numbers between two consecutive integers that represent pieces of the difference between the two numbers. The ratio of two integers is a type of fraction, and there are an infinite number of them between two consecutive integers. Together, the integers and these fractions form the set of rational numbers. The division operation revealed the existence of fractions and expanded our number set to the rational numbers.

Some physical situations produce another type of fraction that is not the ratio of integers and not a rational number.[10] For example, we cannot express the ratio of the circumference of the circular pizza pan to its diameter as the ratio of two integers. These types of ratios led to the discovery of irrational numbers. The number pi (π), which is the value of the circumference-to-diameter ratio, is an irrational number.[11] Like the rational fractions, the irrational numbers fall between integers, and there are an infinite number of them between two consecutive integers. Once again, the division operation revealed the existence of a set of numbers that expands our overall number set.

Together, the rational numbers and the irrational numbers form the set of real numbers. We started with the natural numbers, 1, 2, 3, etc., numbers used to count objects. Then, subtraction revealed the need to add the number zero to our number set in order to describe having no objects present, creating the whole numbers. Subtraction also revealed the need for numbers less than zero, which led to the discovery of negative numbers, and the formation of the set of integers. Dividing large integers into smaller ones revealed fractions, which expanded the number set to the rational numbers. Finally, dividing two non-integer rational numbers produced fractions called irrational numbers that, when combined with the rational numbers, form the set of real numbers.

The real numbers provide complete closure for addition, subtraction and multiplication, and closure for division when the denominator is not zero. By closure, we mean that we can add, subtract and multiply any two real numbers and the result is another real number. We can divide any real number except zero into any other real number and get a real number back. When we divide zero into a number, we call the result singular, infinite or undefined, each term indicating a value outside of the realm of real numbers.[12]

Real numbers allow us to describe practically all the world around us. We use real numbers to analyze the motion of bodies. Financial processes and activities all use real numbers. We design and build structure and devices using real numbers. We count calories, we purchase items, we pay bills, we keep score and we track time, all using real numbers. Most people go through their whole lives never needing more than the real numbers to live.

However, there are mathematical situations for which no real numbers apply. For example, there is at least one operation, the square root operation, for which real numbers do not provide closure. There are two real solutions for the square root of every positive real number, but no real solution for the square root of any of the negative real numbers. The equation $x^2 + a = 0$, or, $x^2 = -a$, has no real solution when a is a real number greater than zero. As in the past, this appears to be a signal that our number system needs expanding. It seems reasonable to expect that a complete number system should close every mathematical operation. Therefore, the set of real numbers does not appear to be a complete set of numbers.

Consider the square operation ($x * x = x^2$). If we let x equal 12, which is the same as +12, then we know that multiplying 12 times 12 gives 144 (12 * 12 = 144). We also know that when we multiply two positive numbers, the resulting number is another positive number (+ * + = +). Therefore, that makes +12 * +12 = +144, or 12 * 12 = 144. The square of +12 is +144, so +12 is a square root of +144 . Now, if we let x equal –12, then, again we know that 12 * 12 = 144. However, now our two 12's are negative. Still, we know that multiplying two negative numbers gives a positive result (– * – = +), so that –12 * –12 = +144. Therefore, the square of –12 is also +144 and –12 is also a square root of +144. Clearly, the only way to get a magnitude of 144 from a square is to use a form of the number 12. However, there are no 12's on the real number line other than +12 and –12. Therefore, there is no form of the number 12 on the real number line that, when squared, gives the number –144.

As the numbers that are less than zero appeared initially, the square root of negative numbers seems to have no physical significance. For the most common instance where squares occur, areas, negative values are nonsense. One cannot have a negative area, so why would one need the root of a negative number? However, this shortcoming revealed that numbers other than real numbers must exist. There is no reason to assume that one number is fundamentally different from any other number. Positive numbers and negative numbers are the same kinds of entities, just at different places on the real number line. Therefore, it seems reasonable to assume that if positive numbers have square roots, then negative numbers must also have square roots.[13]

That 12 is the square root of 144, but no 12 on the real number line gives −144 when squared suggests that another form of 12 must exist. Since this 12 is not on the real number line, it also implies that another number line, in addition to the real number line, must also exist. This realization led to the discovery of the imaginary numbers.[14] The other 12 discussed above is imaginary 12 or 12i. This 12 resides on a number line essentially independent of the real number line called the imaginary number line.

The imaginary number line is similar to the real number line in that it has integers, rational numbers and irrational numbers. It has positive and negative numbers, and it extends out from zero to infinity in both the positive and negative directions. However, the imaginary number line is perpendicular to the real number line. This makes imaginary numbers 90° out of phase with the real numbers. They operate independent of the real numbers. This does not change the nature of imaginary numbers. They are still numbers like real numbers. Their rotation is actually no different from that of negative real numbers, which are 180° out of phase with positive real numbers on the real number line.

To distinguish the imaginary numbers from the real numbers, imaginary numbers carry the imaginary operator i, which is equal to the square root of −1, so that $i^2 = -1$. For example, 20 on the imaginary number line is denoted 20i, and imaginary x is xi. To be consistent, we could give real numbers a real operator, say, r, which would be equal to the square root of 1, so that $r^2 = 1$, and use 20r to denote 20 on the real number line. However, because both r and r^2 are just 1, the r does not modify the number it is attached to. Therefore, real numbers do not need to carry the designation.

The imaginary operator actually serves the same purpose for imaginary numbers as the negative sign (−) does for negative numbers. It rotates the position of the number. As shown in Figure 1.1 below, if moving to the right on the real number line is moving in the positive direction, placing the negative sign on a number says to rotate the number 180° counterclockwise about its origin. Similarly, if a number indicates a position on the real number line, adding the imaginary operator says to rotate the resulting number 90° counterclockwise about its origin. A plus sign or no sign indicates to rotate the number 0°.

In the figure, vectors represent numbers. The positive 12 (+12) vector starts at zero on the real number line and stretches to the right out to 12. Its positive sign then causes it to rotate 0° counterclockwise about its endpoint at 0. Since 0° is no rotation, the vector remains where it is, and ends up pointing to 12 on the real number line.

For the negative 12 (−12), the vector again starts at zero on the real number line and stretches to the right out to 12, as it did for positive 12. Now, however, the negative sign causes the vector to rotate 180° counterclockwise about its endpoint at zero. This causes the tip of the vector to end up at negative 12 on the real number line. Negative 12 is a positive 12 rotated 180° about its origin.

The imaginary 12 (12*i*) vector starts at zero on the real number line and stretches to the right out to 12, as did the other two. It then rotates 90° counterclockwise about its origin at zero. This causes the vector to land perpendicular to the real number line onto the imaginary number line. Now, the vector is pointing at positive 12 on the imaginary number line. Imaginary 12 is a positive real 12 rotated 90° about its origin.

All three numbers start out as 12, then their modifier, +, −, or *i*, rotates them into their final position. A positive number is the positive number rotated 0° about its origin. A negative number is the positive number rotated 180° about its origin. An imaginary number is the positive number rotated 90° about its origin. Figure 1.1 shows that the positive, negative and imaginary numbers all behave consistently.

This way of modeling numbers makes it easy to see why squaring a negative number gives a positive number, and why squaring an imaginary number gives a negative number. Now, for 12 times 12, rotate the first positive 12 zero degrees, which gives 12. Multiply that 12 by 12 for the second 12, which gives 144; and rotate it 0°, which is still 144. Multiplying −12 times −12 becomes positive 12 rotated 180° for the first negative sign, which gives −12. Multiply this times 12, for the second 12, which gives −144. Rotate the −144 another 180° for the second negative sign, which gives 144. Finally, for 12*i* times 12*i*, rotate the first positive 12 counterclockwise 90° for the first *i*, which gives 12*i*. Then, multiply this by 12 for the second 12, which is 144*i*. Rotate the 144*i* another 90° counterclockwise for the second *i*, which gives −144.

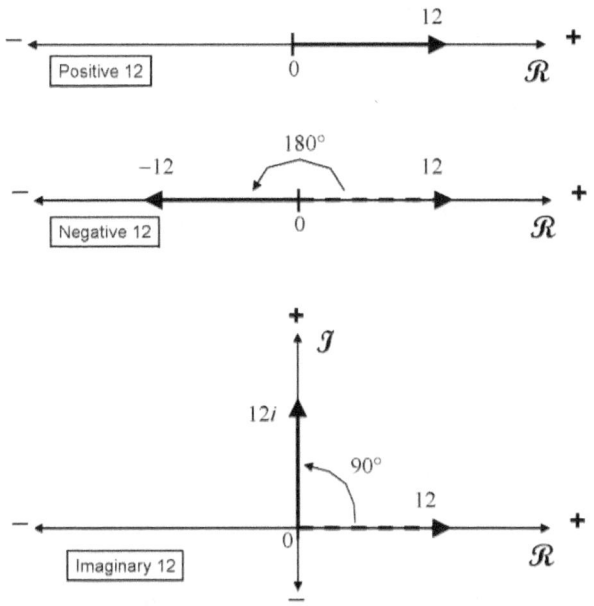

Figure 1.1: Positive, negative and imaginary numbers.

The set of imaginary numbers is a facet of the number system form-ing the structure of the universe, like the integers, the rational numbers and the irrational numbers of the real number set. However, the imagi-nary numbers are not a part of the set of real numbers as the others are. They are a set of numbers independent of the real numbers.

Imaginary numbers rarely ever occur in day-to-day life. Most who know of them consider them abstractions or mathematical devices. However, we eventually discovered that imaginary numbers play im-portant roles in describing some physical phenomena occurring in the real world such as harmonic oscillation, alternating current and wave mechanics. The mathematical expressions for trigonometric functions such as sine and cosine also contain imaginary numbers.[15] Therefore, we need imaginary numbers to describe real world physical phenome-na completely. As with negative numbers before them, the seemingly abstract imaginary numbers do actually have real world applications.

The Physics of Bodies in Real and Imaginary Spaces

This need of imaginary numbers in the real world seems to imply that there is a world greater than the real world made of the real world and an "imaginary" world. The word "imaginary" is an unfortunate word choice for describing that number system. Attempting to contrast the word "real" in real numbers, the label "imaginary" gives the impression that these number do not really exist.[16] In fact, they do exist.

For mathematics, these terms are standards. However, for describing the physical world, let us replace the word "real" with "apparent," and the word "imaginary" with the word "hidden." Now there is an apparent world, A, described by real numbers, and a hidden world, H, described by imaginary numbers. Together, these form what we will call the total world, $\mathscr{T}$.

The total world is the realm through which we are about to travel in order to reevaluate physics. First, we will consider what kind of space must make up the apparent and hidden realms of the total world. Then, discuss mass in the apparent world and how the hidden world aspects of the total world affect it. Next, we will explore the possibility of mass in the hidden world, and how it likely behaves. Finally, we will consider how mass in the hidden world affects the physics of the apparent world. Let us begin our journey.[17]

2
Coordinate Systems

We live in the real world, which, henceforth, we will refer to as the apparent world, and designate with the symbol A. We call it the apparent world because we can readily see it around us. It contains all of the mass and energy we know of. All of the objects we can see, the Sun, the Moon, the stars; plants, animals and people; solids, liquids and gases; all exist in the apparent world. These objects all reside within the space associated with the apparent world. This apparent world space consists of at least three dimensions, typically referred to as length, width and height.[1] We know from experience that objects residing within the apparent space can also move through the space. When we want to describe the positions of objects within this space, we typically use the familiar Cartesian system of coordinates from Euclidian geometry, shown in Figure 2.1.

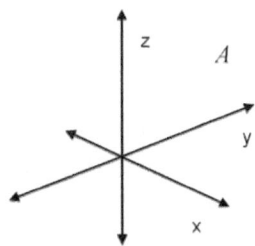

Figure 2.1: Coordinate system of the apparent world.

For apparent space, we define three independent directions, using what we call axes, as the standard for identifying the location of an object in space. Each axis has an origin labeled zero, and extends infinitely out from zero in a direction labeled with positive real values and in the opposite direction with negative real values. Each axis is perpendicular to the other two axes and they all intersect at their origins. We typically label these axes x, y, and z, so that there are three indices or coordinates, the x-coordinate, the y-coordinate and the z-coordinate, that pinpoint the location of an object in apparent space. Each point in apparent world space has a unique (x, y, z) coordinate.[2]

Let us propose that, because a world we call the apparent world associated with the real numbers exists, the existence of imaginary numbers indicates the existence of an "imaginary" world that we call the hidden world, which we designate with the symbol H. We call it the hidden world because, as far as we know, we cannot see into it. It seems reasonable to assume that this hidden world contains objects and physics, just as our apparent world does, and that a space permeates it. There is no reason to believe that space in the hidden world is any different in structure from the space that fills the apparent world.

The Cartesian coordinate system, which works to locate points in our apparent space, should work equally as well for locating points in hidden space. For the hidden world, three perpendicular axes establish the space and three coordinates map any point in the space, as in the apparent world. However, the hidden world axes have imaginary numbers as indices, so that a point in hidden space is denoted (ix, iy, iz). Figure 2.2 shows a diagram of the hidden world coordinate system.

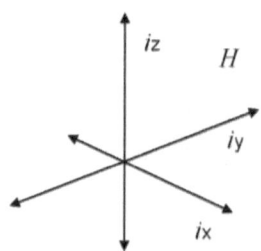

Figure 2.2: Coordinate system of the hidden world.

The existence of the hidden world would mean that it probably contains some form of mass and energy just as our apparent world does. Consequently, objects likely reside in it and things likely happen there. If there is mass and energy in the hidden world, then the mass probably possesses velocity and momentum, and moves through the hidden space. However, from our apparent world, we cannot readily see into the hidden world, ergo, the label hidden. We may not be able to detect anything happening in the hidden world either, unless there is some connection or exchange of information occurring between the two.

Occurrences likely originating in the hidden world do not normally appear to spill over into our apparent world. We do not encounter objects or phenomena in our apparent world that we cannot explain completely with apparent world physics. Well, except, perhaps, why light appears sometimes as waves and sometimes as particles; or, why light has energy but no mass; or, why light travels at a constant velocity regardless of its frame of reference;[3] or, why, in quantum mechanics, particles can exist, disappear, then reappear in a different place. And then there are the even roots of negative numbers. Why, in the real world, do negative numbers appear to have no even roots? The answer is because those roots are in imaginary space. They reside in the hidden world. It seems the apparent world needs something that only the hidden world appears to have, imaginary numbers. Also, multiplying or dividing two numbers in the hidden (imaginary) world produces no solution in hidden space. The solution is a real number residing within apparent space. These ties between problems and solutions in the apparent and hidden worlds suggest that links do exist between the two worlds. Links that may provide us answers to some of this world's perplexing mysteries.

Since there is no way of knowing when a point in the apparent world will need access to hidden world space, and vice versa, every point in the apparent world needs continuous access to the hidden world. Therefore, the two spaces and their corresponding coordinate systems must coexist in a greater "total world", which we shall designate with the symbol $\mathscr{T}$.[4] Coexisting means that they each share a different phase of the same space, but the two are generally inaccessible to each other.

To get an idea of how phases of space might work, consider a large cubic building similar to the one depicted in Figure 2.3. This building represents the total world, which contains all of the space in existence, apparent space and hidden space. There are two doors on adjacent walls of the building. One is at the center of the east side of the building, and the other at the center of the south side of the building. These doors represent the only two ways to enter the building, and consequently, the only two ways to enter the total world.

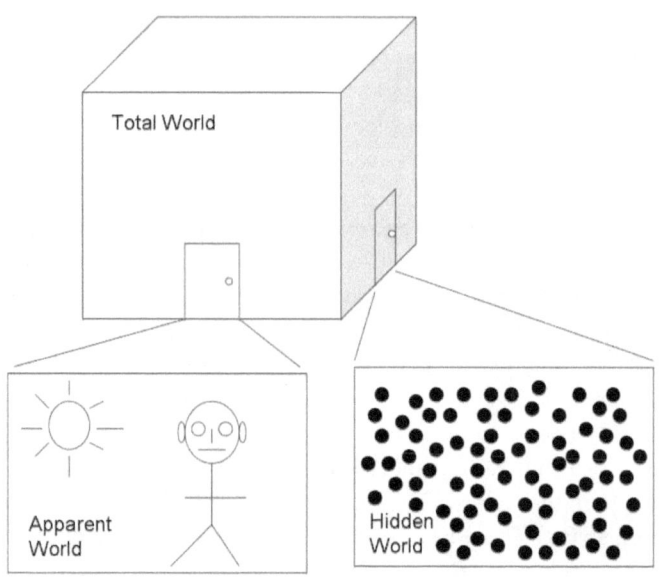

Figure 2.3: Explanation diagram of the total world phase space.

Entering the building through the south side door puts us into the world we currently live in, the apparent world. There, we see people, plants and animals; the Sun rises in the east and sets in the west; there are stars in the sky at night; and all things behave as we have come to know them to. However, entering the building through the east side door puts us into a noticeably different world where the physics is a little bizarre, the hidden world.

In the hidden world world, there may not be any forms of life visible. Instead of stars, and planets, and galaxies, there may only be particles in existence. Both worlds use the same space in the total world building; but we cannot see the hidden world when we are in the apparent world space, and the apparent world is not visible from the hidden world space. Which world we experience depends on how we enter the building. It depends on what phase of the space we reside in.[5]

To remain independent of each other, the hidden world, *H*, must be perpendicular to the apparent world, *A*. The two worlds are 90° out of phase, with only their origins coinciding. This means that the three axes defining apparent space are all perpendicular to the three axes defining hidden space. The *x*-axis of the apparent world is perpendicular to the *ix*-axis, the *iy*-axis and the *iz*-axis of the hidden world. The apparent world *y*-axis is also perpendicular to the *ix*-axis, the *iy*-axis and the *iz*-axis of the hidden world, as is the apparent world *z*-axis.

Figure 2.4 shows this total world configuration. Since the apparent world and the hidden world both have three dimensions, the total world is a six-dimensional space made of two three-dimensional spaces, linked at their origins and perpendicular to each other. The diagram in Figure 2.4 is similar to the Argand diagram in complex numbers, where the horizontal axis of the diagram represents the real numbers, and the vertical axis represents the imaginary numbers. There, the imaginary domain (*y*-axis) is perpendicular to the real domain (*x*-axis). Here, the hidden world, depicted by the *y*-axis, is perpendicular to the apparent world, represented by the *x*-axis.

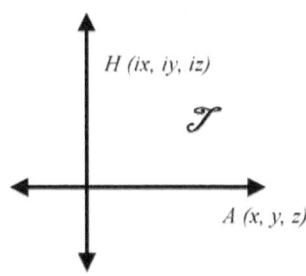

Figure 2.4: Argand-like diagram of the total world.

17

Unlike with complex numbers, the real and imaginary components of entities in the total world do not add together like vectors to produce a magnitude and an angle in the so-called "complex" plane or space. There is no complex space in the total world. Entities exist within either the apparent world or the hidden world, not some "in-between" or "hybrid" world. They may have attributes in both spaces; for example, an apparent velocity and a hidden velocity; but the two velocities remain distinct. They do not form some complex velocity. Because of this, the resultant of combining an apparent world vector and a hidden world vector is the square root of the sum of the apparent world component squared and the hidden world component, including the imaginary operator, squared. So, if the apparent component is x, and the hidden component iy, the resultant is $\sqrt{x^2+(iy)^2}$, which is $\sqrt{x^2-y^2}$, not $\sqrt{x^2+y^2}$, as it is for complex numbers. This results in either a real value or an imaginary value (Figure 2.5). There is no angle $\tan^{-1}(y/x)$ associated with the resultant, so the resultant is not a complex value.[6]

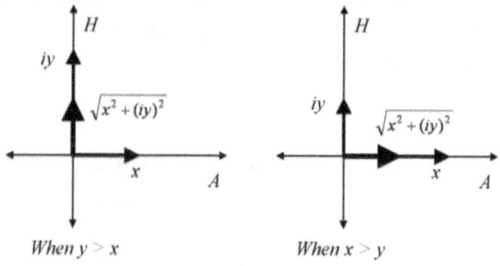

Figure 2.5: Adding vectors in apparent and hidden space.

In the figure above, when the magnitude of the hidden world component is greater than that of the apparent world component, the square of its value is also greater than the square of the apparent world value. The total world resultant is the square root of the difference between the apparent world and hidden world squares. Since the magnitude of the hidden world value that is subtracted from the apparent world value is greater than the apparent world value, the difference is negative. Therefore, the total world quantity, which is now the root of a negative number, is a hidden world value.

Similarly, if the magnitude of the apparent world component is greater than the hidden world component, then the magnitude of the apparent world value squared is greater than the magnitude of the hidden world value squared. This makes the difference of the two a positive value, for which the square root is a positive value. Therefore, the resulting total world value is an apparent world quantity. From the total world perspective, a quantity is either of the apparent world or of the hidden world, not a combination of the two.

Each point in space is a local origin. If in the universe, apparent space and hidden space are infinite, then any point chosen within them can represent the center or origin of them. This means that the origin of apparent space could be the center of the Earth, the center of the Sun, the center of the Milky Way galaxy, or the threshold of the front door of your home. The origin can be anywhere and everywhere. Wherever there is an apparent world origin, by our definition of the total world, there is also a hidden world origin.[7]

If we consider each point in a space a local origin, then each point is the origin for both the hidden space and the apparent space. This means that, as shown in Figure 2.6, each point in the apparent world contains its three Cartesian axes, and the Cartesian axes of a point in the hidden world, and vice versa. Consequently, the hidden world space resides within each point in apparent world space, and the apparent world space resides within each point of the hidden world space. Therefore, each point in a space has three outward dimensions and three inward dimensions.[8]

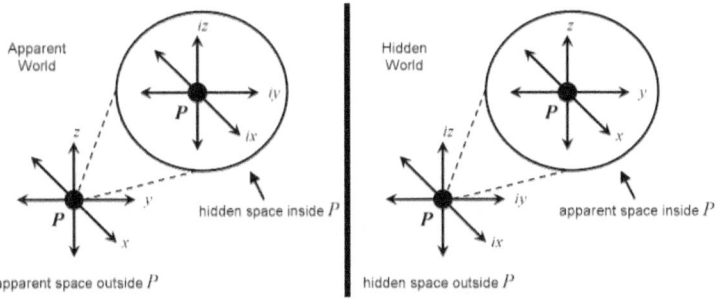

Figure 2.6: Inside and outside of a point in apparent and hidden space.

Because the hidden world is perpendicular to the apparent world, each dimensional axis in the hidden world is perpendicular to each axis in the apparent world. The ix-axis in the hidden world is perpendicular to the x-axis, the y-axis, and the z-axis in the apparent world. The hidden world iy-axis is also perpendicular to the x-axis, the y-axis, and the z-axis in the apparent world, as is the iz-axis in the hidden world. This makes any line drawn in the hidden world perpendicular to every line in the apparent world; and any line drawn in the apparent world is perpendicular to every line in the hidden world.

Although each point in a space is a doorway to the other space, the spaces on either side of the point are independent. In other words, just because something happens at a point in one space does not mean that it happens on the other side of the point in the other space. A line drawn through a set of points in apparent space does not show up as a line drawn through the points in hidden space. Similarly, because a point belongs to an object in apparent space, and that object is moving within apparent space, does not mean that the hidden space side of the point is moving through hidden space. Because the spaces are perpendicular, things can be changing for a point in one space while dormant in the other. This is similar to starting at point (x_1, y_1) on the x-y plane and moving in the x direction, but not the y direction. The x value changes as the point moves, but the y value remains y_1 throughout the motion of the point.

3
Apparent World Mass

The mass we see around us everywhere in our world is apparent world mass. This mass is the stuff that makes up matter, which comes in solid forms such as metals, wood and rocks; liquid forms such as water, oil, and gasoline; and gaseous forms such as air, helium and oxygen. Though they appear outwardly different, just over 100 atoms make up all of the forms of matter we know, and three elementary particles, protons, neutrons and electrons, form all of the atoms. While the electron, and its antiparticle, the positron, appear to be fundamental forms of matter, protons and neutrons are collections of smaller particles that may actually be electrons and positrons.[1] We have yet to observe electrons and positrons break down into smaller particles. They seem to be basic units of apparent mass. However, what substances actually make up the mass forming the electron and the positron are still unknown.

Apparent world mass has volume and takes up space in the apparent world. Due to how the various mass components configure themselves to form matter such as atoms and molecules, mass takes up a lot more space than it has volume. The actual volume of the proton and the electron in a hydrogen atom combined is about 10^{-44} cubic meters. However, the volume of the hydrogen atom is about 10^{-32} cubic meters, 150 trillion times larger.[2] The inflation is due to the forces and fields that compete to pull together and push apart the fundamental particles in the atom. Without these forces and fields, the apparent world would probably be just a large collection unconnected particles.[3]

In the apparent world, when we apply force to a body, the body accelerates. We know that the relationship between the force, the mass of the body, and the resulting acceleration is

$$a = \frac{F}{m},\qquad(3.1)$$

where F is the force, m is the mass and a is the acceleration.[4] A classic example of this is gravity. Near the surface of the Earth, when we release an object above the ground, gravity applies a force to the object, causing it to accelerate toward the ground at about 9.8 m/s[2].

Giving a body momentum in the apparent world gives it velocity. The expression for the velocity, v, a body with apparent mass, m, has due to momentum, p, applied to it is

$$v = \frac{p}{m}.\qquad(3.2)$$

Collisions of billiard balls provide an everyday example of momentum giving an object velocity. When a moving billiard ball collides with a stationary one, it transfers some of its momentum to the stationary ball. In turn, once the momentum transfers, the stationary ball begins to move. The momentum it acquired gives it velocity.

The velocity a body has determines how much kinetic energy it possesses in the apparent world. Kinetic energy is the energy that does work in the apparent world and is force applied over a distance. We can express kinetic energy, E_k, as

$$E_k = \int_0^s F ds,\qquad(3.3)$$

where s is distance. Since the force, F, is just the time derivative of the momentum, or

$$F = \frac{d(mv)}{dt},\qquad(3.4)$$

the expression for kinetic energy becomes

$$E_k = \int_0^s \frac{d(mv)}{dt} ds.\qquad(3.5)$$

The expression ds/dt is just velocity, v, therefore, equation (3.5) is just

$$E_k = \int_0^{mv} v \, d(mv). \qquad (3.6)$$

In classical physics, the mass, m, is constant, so equation (3.6) becomes

$$E_k = m \int_0^v v \, dv = \tfrac{1}{2} m v^2 \Big|_0^v = \tfrac{1}{2} m v^2, \qquad (3.7)$$

yielding the standard classical equation for kinetic energy. However, according to the special theory of relativity, the apparent mass, m, of a body changes as the apparent velocity, v, changes. The relativistic mass expression is

$$m = \frac{m_0}{\sqrt{1 - v^2/c^2}}, \qquad (3.8)$$

where m is the relativistic mass, m_0 is the mass when the body's velocity is zero, and c is the speed of light. Substituting this mass into equation (3.6) gives us

$$E_k = \int_0^v v \, d\left(\frac{m_0 v}{\sqrt{1 - v^2/c^2}} \right). \qquad (3.9)$$

Solving this integral[5] gives relativistic apparent world kinetic energy of

$$E_k = \frac{m_0 c^2}{\sqrt{1 - v^2/c^2}} - m_0 c^2. \qquad (3.10)$$

This kinetic energy expression has two parts: one that is a function of the apparent world velocity, and another due to the speed of light. What is the expression in equation (3.10) telling us about the nature of apparent mass?

From equation (3.10), if we call the first term E_v to indicate that it is the energy due to the apparent world velocity, another expression for the kinetic energy becomes

$$E_k = E_v - m_0 c^2. \qquad (3.11)$$

Equation (3.11) indicates that the kinetic energy of a body, the energy due directly to the body's velocity, is its energy due to its apparent velocity, v, minus a so-called "rest" energy $m_0 c^2$. However, if we write the kinetic energy expression in equation (3.11) as

$$E_k = E_v + m_0(ic)^2 = E_v + E_{ic}, \qquad (3.12)$$

now, the kinetic energy appears to be the sum of an apparent world energy, E_v, and a hidden world energy, E_{ic}. In other words, E_k is actually the total world kinetic energy of the body, and is the sum of the apparent world kinetic energy, E_v, and the hidden world kinetic energy, E_{ic}, which equals $m_0(ic)^2$. The E_v is still an expression of the energy of the body due to its apparent world velocity, v; but the $m_0(ic)^2$ appears to be energy due to a hidden world velocity ic associated with the body. Apparent world bodies appear to have a constant hidden world velocity equal to speed of light, ic, attached to them.

We can get a sense of the physical implications of this expression by expanding the apparent kinetic energy, E_v, into the infinite series[6]

$$E_v = m_0 c^2 (1 + \tfrac{1}{2}\frac{v^2}{c^2} + \tfrac{3}{8}\frac{v^4}{c^4} + \tfrac{5}{16}\frac{v^6}{c^6} + ...), \qquad (3.13)$$

which we can rewrite as

$$E_v = m_0 c^2 + \tfrac{1}{2}m_0 v^2 + m_0 c^2(\tfrac{3}{8}\frac{v^4}{c^4} + \tfrac{5}{16}\frac{v^6}{c^6} + ...). \qquad (3.14)$$

Since the speed of light is very large compared to speeds commonly experienced by masses in the apparent world, the ratio v/c goes to zero for typical apparent world velocities. Therefore, all of the terms inside the parentheses in equation (3.14) go to zero. Consequently, the third term in equation (3.14) goes to zero. This makes the apparent world kinetic energy expression for typical apparent world velocities

$$E_v = m_0 c^2 + \tfrac{1}{2}m_0 v^2. \qquad (3.15)$$

This is the classical kinetic energy plus an $m_0 c^2$. It seems that since the hidden world kinetic energy, $-m_0 c^2$, is a real value that reduces the overall kinetic energy, the apparent world kinetic energy, E_v, has an extra $m_0 c^2$ built into it to offset the hidden world energy.[7]

The kinetic energy of the body, the energy due to its velocity, now appears to be due to its apparent world velocity, v, and (what appears to be) its hidden world velocity, ic. Having this hidden world velocity suggests that all bodies in the apparent world carry a hidden world velocity, ic, equal to the speed of light. This velocity is perpendicular to the apparent world, and therefore, does not cause the bodies to move within the apparent world. Since the bodies only exist within the apparent world and not the hidden world, even though the velocity acts within the hidden world, there is nothing in the hidden world for the velocity to act on. Consequently, nothing moves in the hidden world because of the body's hidden world velocity.[8] Figure 3.1 diagrams the total world velocity of apparent world bodies. The u_v vector is the total world sum of the apparent and hidden velocities.

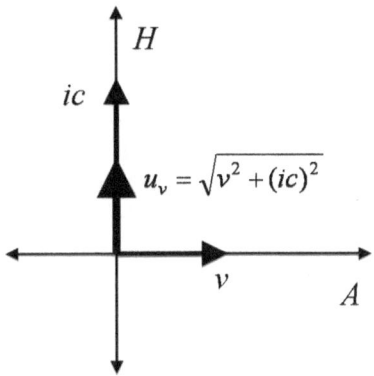

Figure 3.1: **The velocity configuration of an apparent world mass with an apparent world velocity v.**

Because all mass in the apparent world has this hidden world imaginary velocity ic, all mass m_0 in the apparent world has energy equal to $m_0(ic)^2$ or $-m_0c^2$. The minus sign indicates that this is potential energy for apparent world use. It is the energy reserve that the apparent world mass draws from as it gains energy due to its apparent world velocity. If we express the total world energy E_{tw} as

$$E_{tw} = m_0 u_v{}^2,\qquad(3.16)$$

where u_v is the total world velocity, it becomes

$$E_{tw} = m_0\left(\sqrt{v^2 + (ic)^2}\right)^2,\qquad(3.17)$$

or

$$E_{tw} = m_0(v^2 - c^2).\qquad(3.18)$$

From equation 3.18, we can see that when the apparent world velocity is zero, the total world velocity is *ic*, and all of the energy is potential energy. However, as the apparent world velocity of the apparent world mass increases, the total world velocity decreases (becomes a smaller imaginary value) causing the total world energy to decrease. This is because the apparent world mass is using some of the original total world potential energy.

The m_0c^2 energy value is consistent with that derived for mass in special relativity, but arrived at in a much simpler, and easier to understand, way. Now, the mass has energy because it has velocity. In special relativity, mass has this energy, but thought to possess it having no velocity at rest. The apparent world and hidden world configuration of the total world system allows for mass in the apparent world to have velocity, and consequently, energy, without being required to move within either the apparent world or the hidden world.

4
Relativistic Phenomena

In the previous chapter, we determined that all apparent world bodies have an intrinsic hidden world (imaginary) velocity equal to the speed of light. This imaginary velocity is what gives a body its rest energy in the special theory of relativity. Even though it does not cause the body to move in either the apparent or the hidden worlds, it still affects the apparent world body. Since all bodies in the apparent world have this hidden world velocity, when we now consider the velocity of an apparent world body, we must use its total world velocity. This velocity factors in the combined effects of both the apparent world and the hidden world velocities on the body. Let us now see how using this velocity affects various other aspects of the apparent world.

If we give a body with a mass, m, in the apparent world an apparent world velocity, v, its total world velocity becomes the vector sum of this velocity and its hidden world velocity, ic. Since the hidden world velocity is constant and perpendicular to the apparent world velocity regardless of its direction, the total world velocity, u_v, of the body changes only as the apparent world velocity, v, changes and becomes

$$u_v = \sqrt{v^2 + (ic)^2} = \sqrt{v^2 - c^2}. \qquad (4.1)$$

Now, the total world velocity, u_0, at apparent world velocity $v = 0$ equals ic, and as long as the magnitude of v is less than c, the value of u_v will always be an imaginary value less than or equal to ic.[1]

Since the apparent world body has a hidden world velocity, even when it has an apparent world velocity of zero, it has total world momentum. It has this momentum even when it is not moving in the apparent world. The total world is the ultimate closed system. There are no forces outside of it to enter and affect bodies within it. Therefore, the momentum of a body is always conserved in the total world.[2] Consequently, as the apparent world velocity, v, of a body increases, the mass, m, of the body has to decrease relative to what it was at rest in the apparent world to compensate and conserve momentum. In other words, the amount of mass a body has, which we will now denote as m_v, changes as its apparent world velocity changes, so that its momentum in the total world remains constant.[3] This means that we can express conservation of momentum in the total world as

$$m_v u_v = m_0 u_0, \tag{4.2}$$

and we can write

$$m_v = m_0 \frac{u_0}{u_v}. \tag{4.3}$$

The expression in equation (4.3) shows how the mass of a body in the apparent world changes as its apparent world velocity changes in order to keep the total world momentum constant.

Equation (4.3) shows that the apparent world mass changes with the ratio of the body's total world rest velocity to its total world velocity at apparent world velocity, v. We can evaluate the ratio comparing u_0 to u_v, as follows:

$$\frac{u_0}{u_v} = \frac{ic}{\sqrt{v^2 - c^2}} = \frac{1}{\sqrt{\dfrac{v^2 - c^2}{(ic)^2}}} = \frac{1}{\sqrt{1 - v^2/c^2}}. \tag{4.4}$$

This is the Lorentz transformation factor from the special theory of relativity! The ratio of the total world velocity at the apparent world velocity $v = 0$, to the total world velocity at some apparent world velocity v, gives the Lorentz transformation. Using the apparent and the hidden world configurations, its derivation is simple and easily understood as the ratio of the total world velocity of a body when at rest in the apparent world to its total world velocity when moving in the apparent world.[4] Using equation (4.4), now our conservation of momentum expression

$$m_v = m_0 \frac{u_0}{u_v} \qquad (4.5)$$

becomes

$$m_v = \frac{m_0}{\sqrt{1 - v^2/c^2}}, \qquad (4.6)$$

which is the relativistic mass expression from special relativity. Mass changing in relativity is just the total world conserving momentum.

The special theory of relativity was able to show that as the velocity of a body changes, its mass changes. Perhaps, what relativity could not show, because it was only working in the apparent world, is that the relativistic mass expression is an expression of conservation of momentum. As the apparent world velocity of a body changes, the body's mass also changes, offsetting the velocity to maintain the momentum of the body in the total world. Realizing that the Lorentz transformation factor is just this total world velocity ratio, it may help us to gain insight into the true natures of other fundamental relativistic expressions when framed within the total world system.

The special theory of relativity reveals two other phenomena that involve the Lorentz transformation: the Lorentz-Fitzgerald contraction, and time dilation. In the Lorentz-Fitzgerald contraction, a moving object appears to contract along the line of its velocity to an observer standing still. For example, a yardstick that measures 36 inches to someone holding it in a moving car may appear to be only 35 inches to someone standing still outside the car. The person standing still outside the moving car measures a shorter yardstick. The time dilation makes time appear to slow down to a moving clock relative to what an observer standing still sees. In other words, a person inside a moving car might measure 10 seconds for the car to get from point A to point B, but a person standing still outside of the car measures 15 seconds for the trip. Time is passing faster for a person outside of the car standing still than for a person inside the moving car.

In special relativity, both phenomena are due to frame-of-reference differences between a moving observer and an observer standing still.

The velocity of the moving frame of reference supposedly distorts distance for stationary observers relative to observers moving in the frame, and distorts time for the moving observer relative to observers in the stationary frame of reference. However, perhaps the revelation that the Lorentz transformation is just a ratio of total world velocities puts an entirely different spin on what is happening in these two phenomena.

The Lorentz-Fitzgerald contraction from special relativity indicates that a rod appears shorter to a stationary observer when it is moving than when it is at rest. The relativistic equation for this phenomenon is

$$L_v = L_0 \sqrt{1 - v^2 / c^2}, \qquad (4.7)$$

where L_v is the length the rod appears to be to a stationary observer when the rod is moving, and L_0, the length the rod appears to be when it is also stationary. The length changes with the reciprocal of the Lorentz transformation. Since, in the total world system, the Lorentz transformation is just the ratio of the total world velocity at apparent world velocity zero to the total world velocity at apparent world velocity, v, we can write the expression for the Lorentz-Fitzgerald contraction given in equation (4.7) as

$$L_v = L_0 \frac{u_v}{u_0}, \qquad (4.8)$$

or

$$\frac{L_v}{u_v} = \frac{L_0}{u_0}. \qquad (4.9)$$

Equation (4.9) indicates that the length of the rod changes proportionally with the change in total world velocity of the rod. As the apparent world velocity increases, the total world velocity becomes a smaller imaginary value. As that total world velocity gets smaller, to an observer with apparent world velocity zero, so does the apparent length of the rod. Increasing the apparent world velocity of the rod causes its length to contract from the perspective of a stationary observer.

The ratio of the rod's length to its velocity is the amount of time it takes the rod to travel the distance equal to its length ($t = d/v$). Equation (4.9) indicates that, as the rod's total world velocity changes, its length adjusts so that it always takes the same amount of time to travel its length in the total world. Remember, as the apparent world velocity gets higher, the total world velocity – which is imaginary for real objects – gets lower. So, as the rod gains apparent world velocity, it gets shorter because its total world velocity is getting smaller.

For example, consider the length of a rod moving at an apparent world velocity of nine-tenths the speed of light, or $0.9c$, which has a length of 1 meter when its apparent velocity is zero. The total world velocity of the rod is

$$u_v = \sqrt{v^2 - c^2} \, , \qquad (4.10)$$

so that when $v = 0$, the total world velocity of the rod is

$$u_0 = \sqrt{(0)^2 - c^2} = ic \, ; \qquad (4.11)$$

and when $v = 0.9c$, its velocity is

$$u_{0.9c} = \sqrt{(0.9c)^2 - c^2} = 0.436ic \cdot \qquad (4.12)$$

From equation (4.8) we get

$$L_{0.9c} = L_0 \frac{u_{0.9c}}{u_0}, \qquad (4.13)$$

or

$$L_{0.9c} = L_0 (0.436) \cdot \qquad (4.14)$$

Therefore, when the rod is moving at a speed of $0.9c$ in the apparent world, its total world velocity is only 43.6% of what it was when the rod was at rest in the apparent world. Since total world physics says that the length of the rod changes proportionally to the change in the rod's total world velocity, traveling at $v = 0.9c$ in the apparent world, the rod must appear to be only 436 centimeters long to an observer standing still. This is consistent with results from the special theory of relativity.

The other relativistic expression that the total world system sheds some light on is the time dilation. Special relativity shows that, as the velocity of a clock increases, it runs slower relative to a clock that is not moving. In other words, a longer time passes between ticks of the moving clock than for those of the stationary clock. From the perspective of an observer standing still, it takes more than a stationary clock second for a second of the moving clock to elapse. Special relativity expresses the time dilation as

$$t_v = \frac{t_0}{\sqrt{1 - v^2/c^2}},$$ (4.15)

where t_v is the time elapsed between ticks of a clock when the clock is moving, and t_0, the time elapsed between ticks of the clock when it is stationary. Equation (4.15) multiplies the length of time of the stationary clock ticks by the Lorentz transformation to give the length of time of the moving clock ticks. As we did with the relativistic expression for the Lorentz-Fitzgerald contraction, we can replace the Lorentz transformation in equation (4.15) with the ratio of the total world velocities at apparent world velocity zero and apparent world velocity, v. Using the total world expression of the Lorentz transformation, the time dilation expression becomes

$$t_v = t_0 \frac{u_0}{u_v},$$ (4.16)

or

$$t_v u_v = t_0 u_0.$$ (4.17)

This shows that the product of a tick of the clock and its velocity is always the same in the total world. In other words, in the total world, the distance a clock would travel in one of its ticks is always the same regardless of the clock's total world speed.

Remember that as the apparent velocity increases the total world velocity decreases. This means the above expression indicates that as a clock travels faster in the apparent world, the time between ticks of that clock becomes longer relative to a clock traveling slower than it is.

5
Hidden World Mass

Since there is mass in the apparent world, it seems reasonable to assume that there is mass in the hidden world. The hidden world is a three dimensional space as is the apparent world. There is no reason to believe that hidden world space is made of different stuff than apparent world space. Therefore, if apparent world space spawns apparent world mass, then hidden world space should produce hidden world mass. However, some questions about the hidden world mass need answering in order to understand how physics likely works in the hidden world.

For example, what should the hidden world mass look like? Mass in the hidden world may be the same as apparent world mass, but it may be some mysteriously strange entity not known to the apparent world. How does hidden world mass behave within the hidden world? When masses interact within the hidden world, they may follow the same rules as interacting masses in apparent space, but they may act completely different. Does the hidden world mass have any effect on the apparent world? Things may be happening in the apparent world that are the results of things happening in the hidden world that we cannot see. How do we know that there is mass in the hidden world? Unexplained phenomena in the apparent world may indicate the existence of hidden world mass. The following discussion considers these questions and tries to paint a picture of what hidden world mass is.

Even in the apparent world, mass is a somewhat mysterious entity. At the lower level in the apparent world, mass comes in packets as elementary particles such as electrons and positrons. However, we do not know what it is those packets actually contain or how they form.[1] Do they contain concentrated energy? Do they contain wads of space? Do they contain some substance we have yet to discover? We do not know. What we can speculate on is that the existence of the particles is due to some property of apparent space. Apparent space provides whatever ingredients these particles need to form. Therefore, whatever apparent world mass is, hidden world mass is likely similar to it.

Mass in the hidden world probably takes the form of fundamental particles akin to electrons and positrons that contain hidden world stuff. The space in the hidden world should be similar to (if not the same as) the apparent world space, so that it, too, contains the things needed to produce mass particles in it.

The ingredients in the hidden world mass particles are probably the same basic ingredients that make up apparent world mass. However, because it is of the hidden world (which is imaginary space), the hidden world mass is likely not the same as the apparent world mass. Hidden world mass may be a different recipe of the ingredients or a different physical arrangement of them than the apparent world mass.

The hidden world mass is also 90° out of phase with the mass of the apparent world. Mathematically, this makes it imaginary, and we designate it *im*. What this means is that mass from the hidden world probably does not behave in the hidden world or the apparent world as apparent world mass does in the apparent world. For example, if there are fields associated with hidden world mass (gravitational, electric, etc.) as are with apparent world mass, they probably do not act in the apparent world as apparent world fields do. Because of this, it is likely hidden world mass cannot combine with mass in the apparent world to form a contiguous mixture (e.g., hidden world electrons cannot be parts of apparent world atoms). The two types of masses likely retain their identities through separation when brought together, similar to mixing water and oil.

Because the hidden world mass is 90° out of phase with the apparent world, it cannot be seen directly in the apparent world. In other words, if there is hidden world mass in the apparent world, we cannot see it. It is invisible to us in the apparent world. The only way we can tell that hidden world mass is present is through it interacting somehow with the apparent world. If it transfers energy or momentum to apparent world masses, it makes its presence known.

All apparent world mass has a hidden world velocity equal to the speed of light, ic, intrinsic to it. This gives the apparent world mass energy $-mc^2$ when it is not moving in the apparent world. Once again, there is no reason to believe the two worlds are set up differently. Therefore, we will assume that all hidden world mass also has an intrinsic velocity equal to the speed of light, c, but that the velocity is an apparent world velocity. The velocity vector layout for hidden world mass is similar to that of apparent world mass, but now the velocity that changes acts within the hidden world, as shown in Figure 5.1.

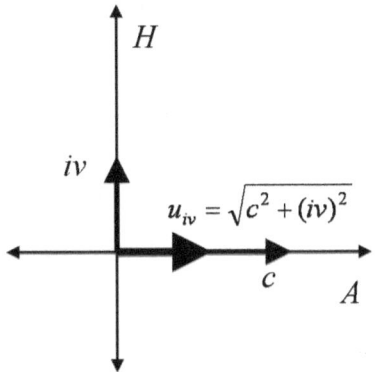

$$u_{iv} = \sqrt{c^2 + (iv)^2}$$

Figure 5.1: The velocity configuration of a hidden world mass with a hidden world velocity *iv*.

In this configuration, the velocity that changes, iv, is a hidden world (imaginary domain) value. It causes the imaginary mass in the hidden world to move within the hidden world space. As long as its magnitude is less than c, the resultant total world velocity, u_{iv}, for the mass will be an apparent (real) velocity. The apparent world velocity component, c, of the hidden world mass is a real domain value and is constant for hidden world mass. Since it acts within the apparent world, this velocity does not cause the mass, im, residing in the hidden world to move. Because of this, hidden world mass always has energy imc^2

when it is not moving in the hidden world; similar to the mc^2 energy apparent world mass has at rest in the apparent world.

Similar to apparent world mass, the hidden world mass only moves in hidden space when it has a hidden world (imaginary) velocity, iv. As the hidden world velocity of hidden world mass increases, the total world velocity of the mass decreases, just as the total world velocity for apparent mass decreases as the apparent world mass gains apparent world velocity. As the hidden world velocity increases, its imaginary value gets larger. When squared, it becomes a larger and larger negative number. Adding it to the square of the real world velocity, c, of the hidden world mass gives the resultant total world velocity. The sum of the two components gets smaller and smaller, making the total world velocity a smaller and smaller real number. For example, in Figure 5.1, if velocity $iv = 0.6ic$; then $(iv)^2 = -0.36c^2$. Since the total world velocity u_{iv} is

$$u_{iv} = \sqrt{c^2 + (iv)^2},$$ (5.1)

then, for $iv = 0.6ic$ it becomes

$$u_{iv} = \sqrt{c^2 - 0.36c^2} = 0.8c.$$ (5.2)

But, if $iv = 0.8ic$, which is a greater hidden world velocity than before; then $(iv)^2 = -0.64c^2$, and the total world velocity becomes

$$u_{iv} = \sqrt{c^2 - 0.64c^2} = 0.6c,$$ (5.3)

which is a smaller total world velocity than before. So, increasing the hidden world velocity of hidden world mass reduces the total world velocity of the mass.

All mass in the hidden world has an intrinsic velocity equal to the speed of light, c, but acting in the apparent world. As with apparent mass, this velocity is perpendicular to the hidden world within which the hidden mass resides, and therefore does not cause the hidden mass to move within the hidden world. Since the mass exists within the hidden world and not the apparent world, it may seem that, even though the velocity acts within the apparent world, there is nothing for the velocity to act upon in the apparent world. However, a subtle difference between the two worlds causes the hidden mass to behave differently than the apparent mass.

When a hidden world mass is moving through the hidden world at some velocity iv, it produces a momentum that is the product of its imaginary (hidden world) mass im, and its imaginary (hidden world) velocity iv. This momentum has a real (apparent world) value of $-mv$. Because of this, the hidden world mass transfers momentum into the apparent world rather than within the hidden world. It does this even though it is not in the apparent world. In transferring momentum to the apparent world, the hidden world masses map themselves into the apparent world, and give indications of their existence that the apparent world can detect. They appear in the apparent world as entities that have momentum, but because their masses are in the hidden (imaginary) world, appear to have no mass. They are phantom masses in the apparent world, packets of momentum that do not appear to have any mass associated with them.[2]

If there were large masses moving through hidden space, then there would be instances of apparent world bodies experiencing large, unexplained momentum increases in the apparent world. Since this does not happen (as far as we know), it suggests that there are no large masses in the hidden world.[3] The physics of the hidden world apparently does not allow fundamental mass particles to collect to form complex masses such as atoms and molecules. This means that all masses in the hidden world are the same. They are the fundamental particles produced in the hidden world. Therefore, in our apparent world, we should look for these phantom masses as small phantom particles, each carrying a small amount of momentum. Small massless particles that interact with mass in the apparent world sound a lot like photons, the particles that make up light and other electromagnetic radiation. We will take a closer look at this observation in the next chapter.

6
Electromagnetic Phenomena

In the last chapter, we determined that hidden world mass (*im*) moving through hidden world space with a hidden world velocity (*iv*) produces apparent world momentum (*−mv*). Because of their real (apparent world) momentums, hidden world masses with hidden world velocities show up in the apparent world as phantom particles. We call them phantom particles because they carry momentum in the apparent world, but have no apparent world mass. Knowing this, how would we expect these particles to behave in the apparent world?

Let us consider a phantom particle in the apparent world. It is the momentum of a hidden world mass projected into the apparent world. Since this mass particle from the hidden world now has a presence in the apparent world, apparent world phenomena affect its apparent world attributes. Now, the apparent world velocity inherent to hidden world masses, *c*, acts on the phantom particle. This causes the phantom particle to move through the apparent world at the speed of light. Therefore, we have a massless phantom particle possessing apparent world momentum moving through the apparent world at the speed of light. This creates a very interesting situation.

The energy of a particle moving at the speed of light is the product of its momentum p, and the speed of light, c, or

$$E = pc. \tag{6.1}$$

For our phantom particle, its momentum is the product of its hidden world mass im and its hidden world velocity iv, or

$$p = (im)(iv) = -mv \cdot \qquad (6.2)$$

Therefore, by moving at the speed of light in the apparent world, this phantom particle carrying momentum $-mv$ has energy

$$E = (im)(iv)c = -mvc \qquad (6.3)$$

in the apparent world. For the particle, both im and c are constants in this energy expression.[1] This means that we can write an expression for the energy of a hidden world phantom particle in the apparent world as

$$E = k_1(iv), \qquad (6.4)$$

where

$$k_1 = imc \cdot \qquad (6.5)$$

This indicates that the phantom particle's energy in the apparent world is directly proportional to the velocity of its source mass in the hidden world. As that velocity gets higher, so does the energy of the phantom particle in the apparent world. Hidden world masses moving through the hidden world at velocity iv, produce phantom particles in the apparent world, whose apparent world energies are proportional to the velocities of those hidden world masses.

While these phantom particles are moving through the apparent world at the speed of light, they are also still moving through the hidden world at velocity iv. We assume, until demonstrated differently, that the hidden world masses move indefinitely along straight lines.[2] Normally, since the hidden world mass is in the hidden world, it moves perpendicular to the apparent world. If we could see it in the apparent world, it would just be a point.[3] However, now that the hidden world mass is projecting onto the apparent world via the phantom particles, these particles trace out a path in the apparent world that corresponds to the movement of the mass in the hidden world. So, the question becomes, how does the mass moving in the hidden world translate into phantom particle movement in apparent space?

The hypothesis here is that the linear motion of the hidden world mass along its straight path in the hidden world translates into a tight circular motion of the phantom particle in apparent space. The phantom particles continuously revolve in fixed circular paths.[4] Figure 6.1 shows how the mapping might occur. Ray *iv* in the hidden world space *H* represents the path of the hidden world mass. Circle *u* in the apparent world space *A* represents the motion of the phantom particle in the apparent world. The figure shows how points a, b, c, and d on the ray map onto the circle. Points a and c coincide on the circle, indicating that one cycle around the circle represents a given distance traveled by the mass in the hidden world. Point d shows another rotation nearly complete. The rate the phantom particle revolves around the circle depends on the speed of the hidden world mass in hidden world space. The higher the hidden world velocity of the hidden world mass, the faster the phantom particle revolves around its circle in the apparent world; or, the higher its frequency of revolution.

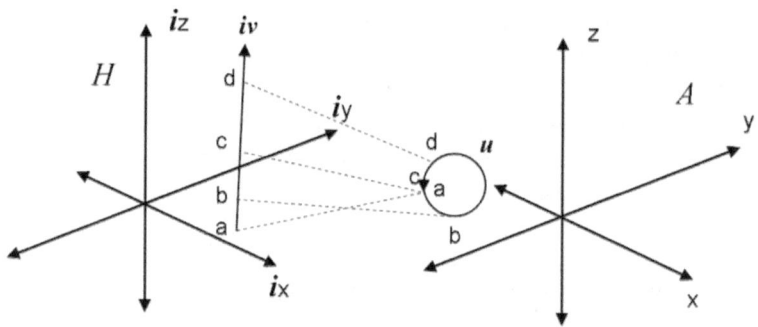

Figure 6.1: Hidden world velocity mapping onto apparent world as a circle.

The velocity, v_r, the phantom particle travels around its circular path is proportional to the velocity, *iv*, of the hidden world mass or

$$v_r = k_2(iv). \tag{6.6}$$

The velocity at which the phantom particle revolves around its circular path is also equal to its frequency of revolution, *f*, times the circumference of the circle, or

$$v_r = 2\pi r f, \tag{6.7}$$

where r is the radius of the circular path. Now, from the last two equations we get

$$k_2(iv) = 2\pi rf \cdot \tag{6.8}$$

If all phantom particles map onto the apparent world with the same size circle,[5] r becomes a constant, so that

$$iv = k_3 f \text{ ,} \tag{6.9}$$

where

$$k_3 = \frac{2\pi r}{k_2} \cdot \tag{6.10}$$

Recalling from equation (6.4) that the energy of the phantom particle in the apparent world is just

$$E = k_1(iv) \text{ ,} \tag{6.11}$$

using (6.9) for iv in this equation gives a phantom particle energy of

$$E = kf \text{ ,} \tag{6.12}$$

where

$$k = k_1 k_3 = \frac{2\pi r(im)c}{k_2} \cdot \tag{6.13}$$

This makes the energy of the phantom particle directly proportional to the particle's frequency.

As mentioned earlier, these phantom particles move at the speed of light in the apparent world because of their apparent world velocity component. Therefore, the particles are moving laterally in the apparent world as they revolve. This combination causes the phantom particles to move in a wave-like motion in apparent world space, with particles that have higher hidden-world velocities having higher wave frequencies in the apparent world. Figure 6.2 depicts a phantom particle moving at the speed of light, c, revolving with a frequency v. The composition of its two motions causes it to follow the wave pattern shown as it moves. The higher the frequency it revolves, the shorter the wavelength of the waves it scribes out. It is likely that, instead of revolving about one axis forming two-dimensional waves as shown in the figure, the phantom particle actually revolves about two axes causing it to scribe out a sphere instead of a circle.[6] Then, the wave pattern would become three-dimensional and the particle would follow a corkscrew pattern or that of a stretched spring as it moved laterally.

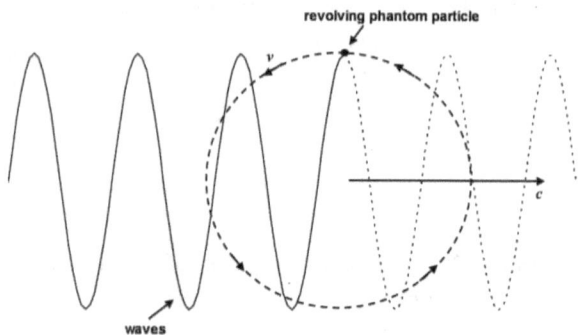

Figure 6.2: Revolving phantom particle moving laterally and creating waves.

Our phantom particles now appear to be particles that have apparent world energy but have no apparent world mass. They travel as waves at the speed of light, and their energy increases as the frequency of the waves increase. The phantom particles appear to be the photons of electromagnetic radiation in the apparent world. In the total world, photons appear to be the projection of hidden world mass particles onto the apparent world through their apparent world momentums.

Now, if we look back at the energy expression we developed for the phantom particles we recall it to be

$$E = kf , \qquad (6.14)$$

where f is the frequency of the wave the phantom particle scribes out and k is a proportionality constant. If our phantom particles are photons, then replacing the k with the constant h, Planck's constant, gives the expression

$$E = hf , \qquad (6.15)$$

which is the quantum physics expression for the energy of electromagnetic radiation, including light. This is a further indication that the phantom particles are the photons of electromagnetic radiation.

The phantom particles get their apparent world velocity from the apparent world velocity associated with the hidden world mass. The only way their apparent world velocities can change is if the apparent world velocities of the hidden world masses generating them change. Light emanating from a source in the apparent world still originates in the hidden world. The apparent world source is merely a container of the source of hidden world mass. The hidden world mass is still in the hidden world. Therefore, because the apparent world is perpendicular to the hidden world, as are velocities originating in the apparent world, applying a velocity to the apparent world source containing the hidden world mass does not add velocity to the hidden world mass.[7]

Consequently, changing the velocity of the source of phantom particles in the apparent world does not affect the velocity of the phantom particles in the apparent world. In other words, the speed of the phantom particles in the apparent world stays the same regardless of their frame of reference. The speed of light is constant in the apparent world because changing the speed of its apparent world source does not change the speed of the hidden world source generating it. This, again, is consistent with the theory of special relativity.

In the hidden world, masses have a constant apparent world velocity c. This velocity produces momentum imc for masses in the hidden world, which are imaginary. They should operate in the hidden world on the hidden world masses.

However, the momentum produced by the apparent world velocity is potential momentum. It is similar to the apparent world rest energy associated with apparent world mass due to its intrinsic hidden world velocity ic. It is momentum that the mass has stored in it, but cannot readily use. Because of this, it does not affect the motion of the hidden mass containing it.

Since the velocities that move masses in the hidden world do not produce hidden world momentum, hidden world mass has no momentum it can transfer to other hidden world masses. Therefore, hidden world masses do not collide with each other, they likely just pass right through each other when they meet.

Notes

[1] I draw this description of scientific evolution primarily from experience and from Thomas Kuhn's book, *The Structure of Scientific Revolutions*. There, early in Section IX he asserts, "*... scientific revolutions are here taken to be those non-cumulative developmental episodes in which an older paradigm is replaced in whole or in part by an incompatible new one.*"

[2] In most references it seems numbers are thought to have been invented; however, I prefer to think of them as having always existed, but not readily apparent to us. Consequently, I say that we discovered them in the same sense that we discovered continents, planets or even physical laws.

[3] Italian mathematician Rafael Bombelli (born 1530) was the first to realize and acknowledge that the square root of negative one is a number.

[4] While this seems obvious, I could not find a reference to support it, so I offer it as speculation.

[5] A set of numbers closes an operation when, using any numbers from the set, the operation gives another number from the set. For example, the addition operation is closed for the set of integers because adding any collection of integers always results in another integer. However, division is not closed for the set of integers because some of the results of dividing integers are fractions, which are not integers.

[6] Originally, people likely counted objects by establishing a one-to-one correspondence between the objects of interest and some counter objects such as pebbles or sticks. The person counting retained one pebble or stick for each item. As long as there was a pebble or stick for each item and an item for each pebble or stick when inventoried, the count was good.

[7] The null or empty set was a recognized concept, but early people did not attach a number to it. Even though the Babylonians invented a symbol for zero around 700 B.C., as late as 500 B.C., Pythagoras and his followers did not recognize zero as a number.

---

[8] While earlier cultures were aware of negative numbers, Pythagoras and his followers did not seem to be aware of them and did not appear to use them.

[9] It seems Pythagoras and his followers resisted the idea of parts of wholes and believed that fractions were multiples of wholes. For example, a pint is not half of a quart; a quart is two pints. The pint is the unit whole.

[10] The story goes that after declaring his famous theorem for the relationship between the legs and hypotenuse of a right triangle, $a^2 + b^2 = c^2$, Pythagoras was horrified to find that he could not prove that when the length of the legs were both one unit, the hypotenuse, $\sqrt{2}$, could not be expressed as the ratio of two whole numbers. This revealed the existence of irrational numbers, numbers that are not the ratio of two whole numbers (which are termed rational numbers).

[11] Some other irrational numbers are e, $\sqrt{2}$ and $\sqrt{3}$.

[12] Division by zero appears to reveal another needed extension to the numbers.

[13] Prior to the 16th century, mathematicians tried, in vain, to extend the real number set to accommodate square roots of negative numbers. They recognized that these roots of negative numbers must exist, but could not come up with a way to extend the real number set to incorporate them. Consequently, they considered them meaningless.

[14] In the 16th century, the Italian mathematician Rafael Bombelli was the first to realize the need for a new number line, the imaginary number line, and suggest that $\sqrt{-1}$ is the number 1 on this number line.

[15] $$\sin(x) = \frac{e^{ix} - e^{-ix}}{2i}, \quad \cos(x) = \frac{e^{ix} + e^{-ix}}{2}.$$

[16] Bombelli labeled the non-real numbers "*imaginary*," casting them in the same derogatory light relative to real numbers, as negative numbers versus positive numbers and irrational numbers versus rational numbers.

[17] The historical references given in the notes above came from Edna E. Kramer's book, *The Nature and Growth of Modern Mathematics*. My copy is a paperback published by Fawcett Premier Books in 1970. The book is still in print and is easily available via the internet.

Notes

Chapter 2

[1] These labels describe the space an object occupies. Length corresponds to the span along the x dimension, width, the span along the y dimension, and height, the span along the z dimension.

[2] We can also express the position of a point in space using the cylindrical coordinates (r, θ, z), where r is the radius, θ is the azimuth, and z is the altitude; or with spherical coordinates (r, θ, φ), where r is the radius, θ is the azimuth, and φ is the inclination.

[3] Light is a very mysterious entity in our apparent world, even though we can characterize much of its behavior.

[4] The total world is essentially the universe. It contains both the apparent world and the hidden world. Some things may happen in the apparent world, and other things may happen in the hidden world, but all things happen in the total world. It is the ultimate closed system.

[5] The basic concept is similar to a toy that used to come in Cracker Jacks® that was a frame that showed one picture if you held it at one angle, but a different picture if you tilted it at another angle.

[6] In complex space, if you have a real component 3, and an imaginary component 4, the magnitude of the complex number vector is 5 ($3^2 + 4^2 = 5^2$), and it is at an angle of 53° ($\tan^{-1}(4/3)$) above the real axis (as shown in Figure 1 below). In the total world, if you have apparent world component 3 and hidden world component 4, the total world value of the two is $\sqrt{-7}$ ($3^2 + 4i^2 = 9 - 16 = -7 = 7i^2$), and the solution vector lies on the hidden world axis (as shown in Figure 2).

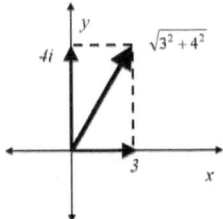

Figure 1   Figure 2

---

[7] We defined the total world as the apparent world and the hidden world joined at the origins; therefore, wherever there is an apparent world point, there is a conjoined hidden world point.

[8] This may be a difficult concept to envision. I am actually suggesting that there is a three dimensional space inside each point of a three dimensional space. The hidden world space inside each apparent world point is the same space, just at a different point within it. In other words, the hidden world space arrived at by entering point (1, 2, 3) in the apparent world is the same hidden space as that arrived at entering point (-7, 23, 49). However, the two points may (or may not) place you at a different point within the hidden world space. This is a different approach to extending the three dimensions of Cartesian space. Typically, extra dimensions are attached as a circle at each point in space, or a sphere or even a torus as Brian Greene describes in chapter 8 of his book *The Elegant Universe*.

Chapter 3

[1] Currently, the fundamental particles that make up protons and neutrons are thought to be quarks and gluons. However, an interpretation of the results of experiments done to probe the proton shows that protons could be made of electrons and positrons. I discuss this in *Infinite Energy Magazine, Vol. 22, Issue 129.*

[2] The proton has a radius of about 10^{-15}m, so its volume is about $4 \times 10^{-45} \text{m}^3$. If the proton is a collection of electrons, then the electron's volume is much less than the proton. So, combined, their volume is still about $4 \times 10^{-45} \text{m}^3$. The radius of the hydrogen atom, which is made of an electron orbiting a proton, is 5×10^{-11}m, so its volume is about $6 \times 10^{-31} \text{m}^3$. Therefore, the volume of the hydrogen atom is about 1.5×10^{14} (150 trillion) times that of the proton and electron combined.

[3] In the atom, the force generated by the electric fields of the proton and the electron cause the two particles to attract each other. However, the orbital velocity of the electron creates centripetal force on the electron that drives it away from the proton. The two forces counter each other, keeping the electron in orbit.

[4] Typically, we are taught Newton's second law of motion as $F = ma$, which suggests that accelerating a mass produces force. I have trouble with the concept of mass spontaneously accelerating. Instead, I believe that the mass accelerates because of a force applied to it. Acceleration is the result of applying force to a mass. For the mass to accelerate, a force must first be introduced.

---

[5] The solution to the integral is as follows:

$$E_k = \int_0^v vd\left(\frac{m_0 v}{\sqrt{1-v^2/c^2}}\right).$$

If we integrate by parts

$$\int x dy = xy - \int y dx,$$

we get

$$E_k = v\left(\frac{m_0 v}{\sqrt{1-v^2/c^2}}\right) - \int_0^v \left(\frac{m_0 v}{\sqrt{1-v^2/c^2}}\right) dv,$$

or

$$E_k = \frac{m_0 v^2}{\sqrt{1-v^2/c^2}} - m_0 \int_0^v \left(\frac{v}{\sqrt{1-v^2/c^2}}\right) dv,$$

which is

$$E_k = \frac{m_0 v^2}{\sqrt{1-v^2/c^2}} - m_0 \int_0^v (1-v^2/c^2)^{-\frac{1}{2}} v dv.$$

If we let $u = 1 - v^2/c^2$, then $du = (-2v/c^2)\, dv$, or, $vdv = -\frac{1}{2}c^2\, du$, and the integral part of the equation above can be written as

$$m_0 \int_0^v (1-v^2/c^2)^{-\frac{1}{2}} v dv = -\frac{m_0 c^2}{2} \int_0^u u^{-\frac{1}{2}}\, du,$$

which becomes

$$-\frac{m_0 c^2}{2} (2u^{-\frac{1}{2}}) \Big|_0^u = -m_0 c^2 \sqrt{1-v^2/c^2}\ \Big|_0^v.$$

Now, the kinetic energy becomes

$$E_k = \frac{m_0 v^2}{\sqrt{1-v^2/c^2}} + \left[m_0 c^2 \sqrt{1-v^2/c^2}\ \Big|_0^v \right],$$

which is

$$E_k = \frac{m_0 v^2}{\sqrt{1-v^2/c^2}} + \left[\frac{m_0 c^2\, (1-v^2/c^2)}{\sqrt{1-v^2/c^2}}\ \Big|_0^v \right],$$

or

$$E_k = \frac{m_0 c^2}{\sqrt{1-v^2/c^2}} - m_0 c^2.$$

[6] The expansion is just $m_0c^2 * f(x)$, where

$$f(x) = (1-x)^{-\frac{1}{2}} - 1 = \sum_{n=1}^{\infty} \frac{f^{(n)}(0)}{n!} x^n,$$

the term $f^{(n)}(0)$ is

$$f^{(n)}(0) = \frac{1*3*5...(2n-1)}{2^n},$$

and

$$x = \frac{v^2}{c^2}.$$

[7] What the m_0c^2 in the E_v term does is ensure that the relativistic total world kinetic energy reduces to the apparent world classical expression at low apparent world velocities ($v \ll c$).

[8] Apparent world velocity only moves apparent world entities and hidden world velocity only moves hidden world entities. Even though the apparent world bodies possess the hidden world velocity, the velocity cannot move them because the hidden world velocity can only move things in the hidden world. However, if, by some chance, the apparent world mass was to enter into the hidden world, its hidden world velocity component should cause it to move through that world at the speed of light. Because of this, the hidden world velocity component of apparent world mass gives the mass potential energy.

Chapter 4

[1] For example, if $v = 0.8c$, then

$$u_v = \sqrt{(0.8c)^2 + (ic)^2} = \sqrt{0.64c^2 - c^2} = \sqrt{-0.36c^2} = 0.6ic.$$

When $v < c$, the value of $u_v < ic$.

[2] Momentum in a closed system is conserved. Absent any outside forces, the momentum of objects interacting within the system remains constant. When the mass in the total world changes its velocity, it does so with impetus from within the total world system, so its momentum within the total world must not change.

[3] How the mass changes (increases or decreases) is somewhat of a mystery, and even though it happens in special relativity theory, I have not been able to find an explanation of what is going on here. This is one of those things that you are

told happens because of the results of the math of the theory, and you know to just accept it. Experiments even appear to validate it, but I do not know of any explanation of where the mass is coming from or going to during the transition.

[4] There are two postulates driving the theory of special relativity: the laws of physics are the same in all frames of reference, and the speed of light is the same for all reference frames. The Lorentz transformation comes out of the need for a transformation between a stationary frame of reference and a moving frame of reference that satisfies these two postulates. The Galilean transformation,

$$x' = x - vt,$$
$$t' = t,$$

where x' is the position in the moving frame, x, the stationary frame, v is the velocity of the body and t is time, fails because, since it assume time is the same in both frames, the speed of light measured is different in the moving frame from that of the stationary frame ($c' = c - v$). In order to remedy this, a factor had to be added to the Galilean transformation that is a function of velocity, so that the transformation becomes

$$x' = k(x - vt).$$

Without showing the math, the time transformation becomes

$$t' = kt + \left(\frac{1-k^2}{kv}\right)x.$$

Since, according to special relativity, both the stationary and the moving frames of reference must measure the same speed of light, then if the stationary frame sees the position of the light change according to

$$x = ct,$$

the moving frame must see the position change according to

$$x' = ct'.$$

Substituting the above expression for x' and t' into the last equation and solving for k gives the Lorentz transformation factor.

Chapter 5

[1] I offer some ideas on what may cause mass to exist in the apparent world in my book *Gravity*.

[2] My thinking here is that when a body is moving, along with its mass, an invisible entity (possibly a type of field) that is its momentum also moves. The fact that momentum is an entity of some sorts allows some or all of it to move from one body to another when to bodies collide.

[3] If there were hidden world buses or planets, for example, made of hidden world mass moving with hidden world velocities, they would create phantom masses projecting large amounts of apparent world momentum into the apparent world. Apparent world masses encountering these would be significantly shaken, appearing to be jarred for no reason in the apparent world.

Chapter 6

[1] Since the apparent world velocity, c, does not move the hidden world mass, im, as long as the hidden world velocity, iv, is not approaching the speed of light, the mass is not relativistic.

[2] The hidden world particles moving through hidden space cannot transfer momentum between themselves, so it seems that interactions between particles should not alter the paths of the particles. Perhaps the particles just pass right through each other when they meet.

[3] Think of a point moving in the x direction, away from point $(0, 3)$ on the y-axis along the line $y = 3$. If you are standing on point $(0, 3)$, all you can see is the point. You cannot see the line that the point traces out as it moves in the x direction that someone standing on, say, point $(2, 0)$ can, because the point is moving perpendicular to the axis that you are standing on. This is the same situation. The hidden world is perpendicular to the apparent world we are standing in, so as the particle moves anywhere in hidden space, we cannot see its path, all we see is a point.

[4] I propose this because lines in one space transform into circles in other spaces in conformal mapping from complex variables. I am suggesting that going from $H(ix, iy, iz)$ to $A(x, y, z)$ in the total world is similar mathematically to going from $w(u, v)$ to $z(x, y)$ in complex space.

---

[5] In Chapter 5, we established that the hidden world mass particles are likely all the same size since we do not appear to encounter large unexplained momentum transfers to bodies in the apparent world. If that is the case, the thinking is that same-sized mass particles in the hidden world scribe out same-sized circular paths in the apparent world.

[6] If we spin the two-dimensional loop making the waves about the line labeled c in Figure 6.2, the result would be a three-dimensional corkscrew wave.

[7] If you have a flashlight, even though the light appears to emanate from the bulb, which is in the apparent world, it is actually coming from the hidden world mass particles in the hidden world. It moves at the speed of light because of the apparent world velocity, c, that the hidden world mass has. The only way to make the light go faster than c, is to increase the apparent world velocity that the hidden world mass has. Changing the speed of the flashlight will not do this.

Index

Index

ABOUT THE AUTHOR

William Stubbs is a retired engineer who independently researches a variety of subjects including physics. He earned a degree in Nuclear Engineering from the University of Tennessee and worked for both private and public engineering organizations during his career. His former employers include General Electric, Westinghouse Electric, the Tennessee Valley Authority, and the U.S. Department of Energy. He has published several articles on physics and nuclear science since retiring in 2005; and self-published the books, *Nuclear Alternative: Redesigning Our Model of the Structure of Matter* in 2008, *Gravity* in 2012 and *Proton Structure* in 2015. In addition to his research, William enjoys listening to, performing, and composing music; developing simple Android apps; solving math and logic puzzles, and spending time with family and friends. He lives in Port St. Lucie, Florida.